Mathstart™
TIME LINES

Get Up and Go!

by Stuart J. Murphy • illustrated by Diane Greenseid

HarperCollins*Publishers*

LEVEL
2

11-06
15·00

To Babci and Dzia-Dzia—
who love to give kids their get-up-and-go
—S.J.M.

To Ida and Rosie,
Queen and Queen of the Doggy Brigade
—D.G.

The illustrations in this book were done with acrylics on Arches watercolor paper.

HarperCollins®, ☖®, and MathStart™ are trademarks of HarperCollins Publishers Inc.

For more information about the MathStart series, please write to
HarperCollins Children's Books, 10 East 53rd Street, New York, NY 10022.

Bugs incorporated in the MathStart series design were painted by Jon Buller.

Library of Congress Cataloging-in-Publication Data
Murphy, Stuart J., date
   Get up and go! / by Stuart J. Murphy ; illustrated by Diane Greenseid.
      p.   cm.   (MathStart)
   "Level 2."
   Summary: Explains through the use of rhyme the concepts of time lines
and addition as a girl gets ready for school with the help of her smart dog.
   ISBN 0-06-025881-0. — ISBN 0-06-025882-9 (lib. bdg.)
   ISBN 0-06-446704-X (pbk.)
   1. Addition—Juvenile literature.   [1. Addition.]   I. Greenseid, Diane, ill.
II. Title.   III. Series.
QA115.M87   1996   95-4736
513.2'11—dc20   CIP
                                                              AC

1 2 3 4 5 6 7 8 9 10
❖
First Edition

Get Up and Go!

You're always so slow.
Let's get up and go!

5

Just 5 minutes more
to snuggle with Teddy.

If you don't get up
you'll never be ready.

7

A *3*-minute stop—
that's all I'll take.

I'd better see
how much time that will make.

9

She's already late—so I'd better try
to keep careful track of the time going by.

I can show her 5-minute snuggle with Teddy like this.

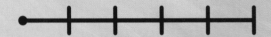

3 minutes to wash looks like this.

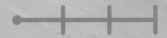

Now I'll put my lines together.

How many minutes have gone by so far?

Then 8 minutes to eat—
I like breakfast the most.

12

I only wish
that you'd toss me some toast.

13

*Now 2 minutes extra
to give Sammie a treat.*

14

Dog snacks are great.
I'm ready to eat.

She's going upstairs and still has a lot to do.
I'd better keep track of these minutes, too.

I'll show the 8 minutes she took to eat breakfast.

Then 2 minutes to give me a treat looks like this.

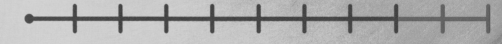

I'll put these lines together. How many minutes have gone by?

Now I'll add this line to my first line.

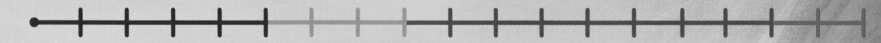

How many minutes have gone by now?

16

17

Then 6 minutes to brush—
both my teeth and my hair.

18

You're still running late
but I'm sure you don't care.

19

And **7** to dress.
That's all I need.

20

Unless you play games . . .
or sit down to read.

21

She's taking so long. I'm never sure why.
I'd better check how much time has gone by.

6 minutes to brush looks like this.

And I can show 7 minutes to dress like this.

Now I'll put them together.
How many minutes do we have now?

Then I'll put all my lines together.
How many minutes have gone by in all?

Now 4 minutes to pack
all the things I can find.

Make sure that you don't
leave your homework behind!

*A **1**-minute hug
and I'll be out the door.*

26

I wish you had time
for just one hug more.

She's finally on her way—she was almost too late.
Now that she's off, everything will be great.

I'll show 4 minutes to pack.

Then I'll show my 1-minute hug.

If I put them together, they'll look like this.
How many minutes do we have now?

Next I'll put all the lines together.
Now how many minutes have gone by in all?

Now I know how much time she took to get ready,
from the time she woke up and snuggled with Teddy.

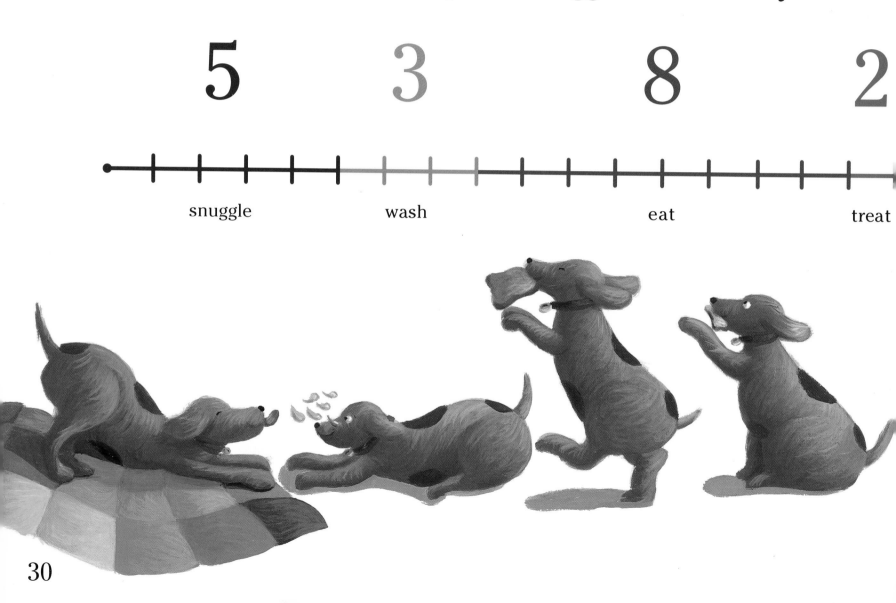

5   3   8   2

snuggle   wash   eat   treat

She took 5 and 3, 8 and 2, 6 and 7, 4 and 1.
That's 36 minutes—and my work was done!

6   7   4   1

brush   dress   pack   hug

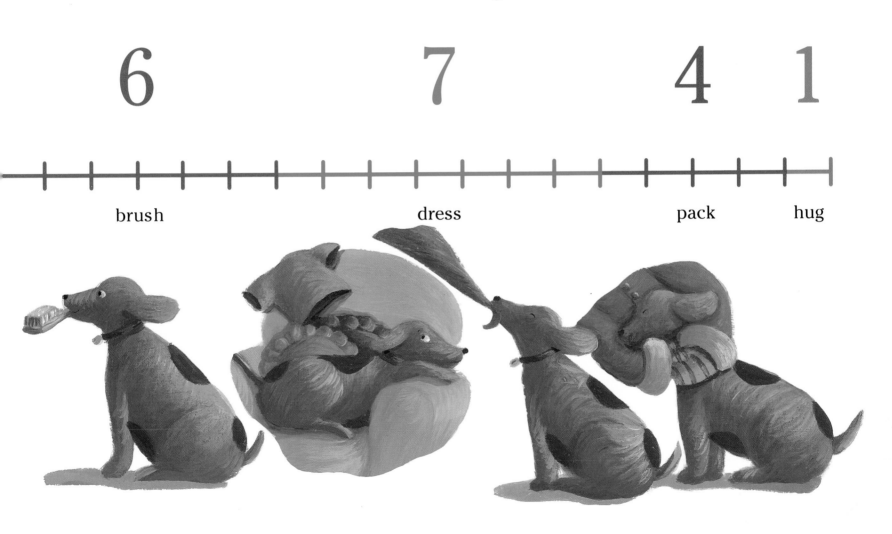

Now she's off to school, and I'm feeling fine.
The rest of the day is totally mine!

If you would like to have more fun with the math concepts presented in *Get Up and Go!*, here are a few suggestions:

• Read the story together and ask the child to describe what is going on in each picture.

• Ask questions throughout the story, such as "How much time does it take the girl to eat her breakfast?" and "How much time does it take her to brush her teeth?"

• Encourage the child to tell the story using the math vocabulary: "time," "minutes," "plus," and "equals." Talk about which activities take more time and which take less time. How can you tell which take more time by looking at Sammie's time lines?

• Together, draw and color pictures of the child's own morning routine. Time the minutes needed for each activity and use strips of paper, string, or yarn to create a personal time line. Tape the pictures to the appropriate line segments.

• Create time lines for activities that occur at other times of the day. For example, time the sequence of events involved in dinner, such as preparing the food, setting the table, eating, and cleaning up. Which activity takes the most time? Which takes the least time?

• Talk about everyday places you visit—sandwich shops, barber shops, grocery stores— and identify the sequences of events that occur there. Together, create time lines for these situations.

(continued on next page)

33

ollowing are some activities that will help you extend the concepts presented in *Get Up and Go!* into a child's everyday life.

Cooking: Pick a favorite snack to make—for example, a peanut butter and jelly sandwich, nachos, or lemonade—and create a time line of the steps involved in preparing it. Are the steps in the right order? Does the time line show which steps take the most time?

Games: Give each family member a long piece of yarn. Ask them to cut each piece in half to represent daytime and nighttime, and then to cut each half into lengths that represent their activities. Compare lengths. Who sleeps the longest? Who spends the most time eating?

Making plans: Plan a party that will take place from 2:00 to 4:00. What has to be done beforehand? What activities will take place during the party? What has to be done after the party is over? Make a time line of these activities.

he following books include some of the same concepts that are presented in *Get Up and Go!*:

- THE STOPWATCH by David Lloyd and Penny Dale

- FROGGY GETS DRESSED by Jonathan London

- SUNSHINE by Jan Ormerod